JULY 02

Using DNA to Solve Cold Cases

NCJ 194197

Sarah V. Hart
Director
National Institute of Justice

This document is not intended to create, does not create, and may not be relied upon to create any rights, substantive or procedural, enforceable at law by any party in any matter civil or criminal.

Findings and conclusions of the research reported here are those of the authors and do not reflect the official position or policies of the U.S. Department of Justice.

The National Institute of Justice is a component of the Office of Justice Programs, which also includes the Bureau of Justice Assistance, the Bureau of Justice Statistics, the Office of Juvenile Justice and Delinquency Prevention, and the Office for Victims of Crime.

National Commission on the Future of DNA Evidence

In 1995, the National Institute of Justice (NIJ) began research that would attempt to identify how often DNA had exonerated wrongfully convicted defendants. After extensive study, NIJ published the report *Convicted by Juries, Exonerated by Science: Case Studies in the Use of DNA Evidence to Establish Innocence After Trial,* which presents case studies of 28 inmates for whom DNA analysis was exculpatory.

On learning of the breadth and scope of the issues related to forensic DNA, the Attorney General asked NIJ to establish the National Commission on the Future of DNA Evidence as a means to examine the most effective use of DNA in the criminal justice system. The Commission was appointed by the NIJ Director and represented the broad spectrum of the criminal justice system. Chaired by the Honorable Shirley S. Abrahamson, Chief Justice of the Wisconsin Supreme Court, the Commission consisted of representatives from the prosecution, the defense bar, law enforcement, the scientific community, the medical examiner community, academia, and victims' rights organizations.

The Commission's charge was to submit recommendations to the Attorney General that will help ensure the best use of DNA as a crimefighting tool and foster its use throughout the entire criminal justice system. Other focal areas for the Commission's consideration included crime scene investigation and evidence collection, laboratory funding, legal issues, and research and development. The Commission's working groups, consisting of commissioners and other experts, researched and examined various topics and reported back to the Commission. The working groups' reports were submitted to the full Commission for approval, amendment, or further discussion and provided the Commission with background for its recommendations to the Attorney General.

By nature of its representative composition and its use of numerous working groups, the Commission received valuable input from all areas of the criminal justice system. The broad scope of that input enabled the Commission to develop recommendations that both maximize the investigative value of the technology and address the issues raised by its application.

Commission members

Chair

The Honorable Shirley S. Abrahamson
Chief Justice
Wisconsin Supreme Court

Members

Dwight E. Adams
Director
Federal Bureau of Investigation Laboratory

SPECIAL REPORT / JULY 02

Jan S. Bashinski
Chief
Bureau of Forensic Services
California Department of Justice
Sacramento, California

George W. Clarke
Deputy District Attorney
San Diego, California

James F. Crow
Professor
Department of Genetics
University of Wisconsin

Lloyd N. Cutler
Wilmer, Cutler & Pickering
Washington, D.C.

Joseph H. Davis
Former Director
Miami-Dade Medical Examiner
    Department

Paul B. Ferrara
Director
Division of Forensic Sciences
Commonwealth of Virginia

Norman Gahn
Assistant District Attorney
Milwaukee County, Wisconsin

Terrance W. Gainer
Executive Assistant Chief
Metropolitan Police Department
Washington, D.C.

Terry G. Hillard
Superintendent of Police
Chicago Police Department
Chicago, Illinois

Aaron D. Kennard
Sheriff
Salt Lake County, Utah

Philip Reilly
Interleukin Genetics
Waltham, Massachusetts

Ronald S. Reinstein
Associate Presiding Judge
Superior Court of Arizona
Maricopa County, Arizona

Darrell L. Sanders
Chief
Frankfort Police Department
Frankfort, Illinois

Barry C. Scheck
Professor
Cardozo Law School
New York, New York

Michael Smith
Professor
University of Wisconsin Law School

Jeffrey E. Thoma
Public Defender
Mendocino County, California

Kathryn M. Turman
Director
Office for Victim Assistance
Federal Bureau of Investigation

William Webster
Milbank, Tweed, Hadley & McCloy
Washington, D.C.

James R. Wooley
Baker & Hostetler
Cleveland, Ohio

Commission staff

Christopher H. Asplen
Executive Director

Lisa Forman
Deputy Director

Robin W. Jones
Executive Assistant

USING DNA TO SOLVE COLD CASES

Crime Scene Investigation Working Group

The Crime Scene Investigation Working Group is a multidisciplinary group of criminal justice professionals from across the United States who represent both urban and rural jurisdictions. Working group members and contributors were recommended and selected for their experience in the area of criminal investigation and evidence collection from the standpoints of law enforcement, prosecution, defense, the forensic laboratory, and victim assistance.

DNA has proven to be a powerful tool in the fight against crime. DNA evidence can identify suspects, convict the guilty, and exonerate the innocent. Throughout the Nation, criminal justice professionals are discovering that advancements in DNA technology are breathing new life into old, cold, or unsolved criminal cases. Evidence that was previously unsuitable for DNA testing because a biological sample was too small or degraded may now yield a DNA profile. Development of the Combined DNA Index System (CODIS) at the State and national levels enables law enforcement to aid investigations by effectively and efficiently identifying suspects and linking serial crimes to each other. The National Commission on the Future of DNA Evidence made clear, however, that we must dedicate more resources to empower law enforcement to use this technology quickly and effectively.

Using DNA to Solve Cold Cases is intended for use by law enforcement and other criminal justice professionals who have the responsibility for reviewing and investigating unsolved cases. This report will provide basic information to assist agencies in the complex process of case review with a specific emphasis on using DNA evidence to solve previously unsolvable crimes. Although DNA is not the only forensic tool that can be valuable to unsolved case investigations, advancements in DNA technology and the success of DNA database systems have inspired law enforcement agencies throughout the country to reevaluate cold cases for DNA evidence. As law enforcement professionals progress through investigations, however, they should keep in mind the array of other technology advancements, such as improved ballistics and fingerprint databases, which may substantially advance a case beyond its original level.

Chair

Terrance W. Gainer
Executive Assistant Chief
Metropolitan Police Department
Washington, D.C.

Members

Susan Ballou
Office of Law Enforcement Standards
National Institute of Standards and
   Technology
Gaithersburg, Maryland

Jan S. Bashinski
Chief
Bureau of Forensic Services
California Department of Justice
Sacramento, California

Sue Brown
INOVA Fairfax Hospital
SANE Program
Falls Church, Virginia

Lee Colwell
Director
Criminal Justice Institute
University of Arkansas System
Little Rock, Arkansas

Thomas J. Cronin
Chief
City of Coeur d'Alene Police Department
Coeur d'Alene, Idaho

Terry G. Hillard
Superintendent of Police
Chicago Police Department
Chicago, Illinois

Mark Johnsey
Master Sergeant (Ret.)
Division of Forensic Services
Illinois State Police Department
Springfield, Illinois

Christopher Plourd
Attorney at Law
San Diego, California

Darrell L. Sanders
Chief
Frankfort Police Department
Frankfort, Illinois

Clay Strange
Assistant District Attorney
Travis County District Attorney's Office
Austin, Texas

Contributors

Cheryl May
Assistant Director
Forensic Sciences Education Center
Little Rock, Arkansas

William McIntyre
Detective Sergeant (Ret.)
Atlantic County Prosecutor's Office
Homicide Unit
Hammonton, New Jersey

Contents

National Commission on the Future of DNA Evidence iii

Introduction 1

The Long and Short of DNA 5

How Can DNA Databases Aid Investigations? 9

Practical Considerations 13

Identifying, Analyzing, and Prioritizing Cases 17

Introduction

In 1990, a series of brutal attacks on elderly victims occurred in Goldsboro, North Carolina, by an unknown individual dubbed the "Night Stalker." During one such attack in March, an elderly woman was brutally raped and almost murdered. Her daughter's early arrival home was the only thing that saved the woman's life. The suspect fled, leaving behind materials intended to burn the residence and the victim in an attempt to conceal the crime. In July 1990, another elderly woman was brutally raped and murdered in her home. Three months later, a third elderly woman was raped and stabbed to death. Her husband was also murdered. Their house was burned in an attempt to cover up the crime, but fire/rescue personnel pulled the bodies from the house before it was engulfed in flames.

When DNA analysis was conducted on biological evidence collected from vaginal swabs from each victim, authorities concluded that the same perpetrator had committed all three crimes. However, there was no suspect.

For 10 years, both the Goldsboro Police Department and the crime laboratory refused to forget about these cases. With funding from the National Institute of Justice, the crime laboratory retested the biological evidence in all three cases with newer DNA technology and entered the DNA profiles into North Carolina's DNA database. This would allow the DNA profile developed from the crime scene evidence to be compared to thousands of convicted offender profiles already in the database.

In April 2001, a "cold hit" was made to the perpetrator's convicted offender DNA profile in the database. The perpetrator had been convicted of shooting into an occupied dwelling, an offense that requires inclusion in the North Carolina DNA database. The suspect was brought into custody for questioning and was served with a search warrant to obtain a sample of his blood. That sample was analyzed and compared to the crime scene evidence, thereby confirming the DNA database match. When confronted with the DNA evidence, the suspect confessed to all three crimes.

Mark Nelson, special agent in charge of the North Carolina State Crime Laboratory, said, "Even though these terrible crimes occurred more than 10 years ago, we never gave up hope of solving them one day."

Every law enforcement department throughout the country has unsolved cases that could be solved through recent advancements in DNA technology. Today, investigators who understand which evidence may yield a DNA profile can identify a suspect in ways previously seen only on television. Evidence invisible to the naked eye can be the key to solving a residential burglary, sexual assault, or murder. The saliva on the stamp of a stalker's threatening letter, the perspiration on a rapist's mask, or the skin cells shed on the ligature of a strangled child may hold the key to solving a crime.

In Austin, Texas, for example, an investigator knowledgeable about DNA technology was able to solve the rape of a local college student. Having read about the potential for obtaining DNA evidence from the ligature used to strangle a victim, the investigator requested DNA testing on the phone cord used to choke the victim in his case. He realized that in the course of

SPECIAL REPORT / JULY 02

choking someone, enough force and friction is applied to the rope or cord that the perpetrator's skins cells may rub off his hands and be left on the ligature.

The investigator's request paid off in an unanticipated way. In spite of the attacker's attempt to avoid identification through DNA evidence by wearing both a condom and rubber gloves, a reliable DNA profile was developed from the evidence. During the struggle, the attacker was forced to use one hand to hold the victim down, leaving only one hand to pull the phone cord tight. The attacker had to grab the remaining end of the cord with his mouth, thereby depositing his saliva on the cord. Although the developed profile came from saliva rather than skin, DNA not only solved the case in Austin, but also linked the perpetrator to a similar sexual assault in Waco.

Without the investigator's understanding of DNA technology and where DNA might be found, the case may have gone unsolved. The successful review and investigation of unsolved cases require the same basic elements as the investigation of new cases: cooperation among law enforcement, the crime laboratory, and the prosecutor's office. Investigators should be aware of technological advances in DNA testing that may yield profiles where previous testing was not performed or was unsuccessful. The crime laboratory can be essential to the preliminary review of unsolved cases, for example, by providing investigators with laboratory reports from previous testing and consultation regarding the investigative value of new DNA analysis techniques and DNA database search capabilities. Additionally, the prosecutor's office should be involved as soon as a case is reopened so that legal issues are addressed appropriately. It is also extremely important that case reconstruction considers the victim or victim's family and the importance of finality to closing a case.

The successful review and investigation of unsolved cases require cooperation among law enforcement, the crime laboratory, and the prosecutor's office.

Although DNA is not the only forensic tool available for the investigation of unsolved cases, advancements in DNA testing and the success of DNA database systems have inspired law enforcement agencies throughout the country to reevaluate cases previously thought unsolvable. The purpose of this report is to provide law enforcement with a practical resource for the review of old, cold, or unsolved cases that may be solved through DNA technology and DNA databases. "The Long and Short of DNA" and "How Can DNA Databases Aid Investigations?" will educate the reader about the science and technology of DNA testing and DNA databases. "Practical Considerations" provides important background information on legal and practical considerations regarding the application of DNA technology to old, cold, or unsolved cases. Finally, a step-by-step process is provided to help investigators select cases that would most likely be solved with DNA evidence. As investigators advance through this process, they should also keep in mind the array of other technology advancements, such as improved ballistics and fingerprint databases, that may benefit their investigation.

Advancements in DNA technology

Advancements in DNA analysis, together with computer technology and the Combined DNA Index System (CODIS),[1] have created a powerful crimefighting tool for law enforcement. CODIS is a computer network that connects forensic DNA laboratories at the local, State, and national levels. DNA database systems that use CODIS contain two main criminal indexes and a missing persons index. When a DNA profile is developed from crime scene evidence and entered into the forensic (crime scene) index of CODIS, the database software searches thousands of convicted offender DNA profiles (contained in the offender index) of individuals convicted of offenses such as rape and murder. Similar to the Automated Fingerprint Identification System (AFIS), CODIS can aid investigations by efficiently comparing a DNA profile generated from biological evidence left at a crime scene against convicted offender DNA profiles and forensic evidence from other cases contained in CODIS. CODIS can also aid investigations by searching the missing persons index, which contains DNA profiles of unidentified remains and DNA profiles of relatives of those who are missing. Because of the recidivistic nature of violent offenders, the power of a DNA database system is evident not only in the success of solving crimes previously thought unsolvable, but perhaps more importantly, through the *prevention* of crime.

When properly documented, collected, and stored, biological evidence can be analyzed to produce a reliable DNA profile years, even decades, after it is collected. Just as evidence collected from a crime that occurred yesterday can be analyzed for DNA, today evidence from an old rape kit, bloody shirt, or stained bedclothes may contain a valuable DNA profile. These new analysis techniques, in combination with an evolving database system, make a powerful argument for the reevaluation of unsolved crimes for potential DNA evidence.

Knowledgeable law enforcement officers are taking advantage of powerful DNA analysis techniques by investigating crime scenes with a keener eye toward biological evidence. The same new approach being applied to crime scene processing and current case investigation can be applied to older unsolved cases. Law enforcement agencies across the country are establishing cold-case squads to systematically review old cases for DNA and other new leads. This report will serve as a resource to assist law enforcement with maximizing the potential of DNA evidence in unsolved cases by covering the basics of DNA analysis and its application to forensic casework. The report will also demonstrate how DNA database systems, advancing technology, and cooperative efforts can enhance unsolved case investigative techniques.

New laws

Advancements in DNA technology have led to significant changes in many States' statutes, which may affect the manner in which unsolved cases are investigated, filed, and prosecuted. Advancements in the technology have been so significant that laws are being created, amended, and even repealed to take advantage of its ability to identify and convict the guilty and exonerate the innocent. Laws regarding DNA admissibility in court, its use in post-conviction appeals, the creation and expansion of databases, and the extension or elimination of statutes of limitation are examples of the quickly evolving impact of DNA on the criminal justice system. Given the legal changes occurring throughout the country, constant contact and consultation with the local prosecutor is critical not only for the investigation of older cases but for all cases in which DNA may be relevant evidence.

Statutes of limitation

Statutes of limitation may be one of the most difficult issues to overcome when examining older cases. Statutes of limitation establish time limits under which criminal charges can be filed for a particular offense. These statutes are rooted in the protection of individuals from the use of evidence that becomes less reliable over time. For example, witnesses' memories fade as time goes by. However, although some evidence, such as eyewitness accounts, can lose credibility over time, DNA evidence has the power to determine truth 10, 15, even 20 years

The power of a DNA database system is evident not only in the success of solving crimes previously thought unsolvable, but through the prevention *of crime.*

SPECIAL REPORT / JULY 02

The reliability of DNA technology may necessitate the reevaluation of statutes of limitation.

after an offense is committed. States are beginning to realize that the reliability of DNA technology may necessitate the reevaluation of statutes of limitation in the filing of cases.

Database expansion

The use of DNA evidence and convicted offender DNA databases has expanded significantly since the first U.S. DNA database was created in 1989. Although State and local DNA databases established in the early 1990s contained only DNA profiles from convicted murderers and sex offenders, the undeniable success of DNA databases has resulted in a national trend toward database expansion. All States require at least some convicted offenders to provide a DNA sample to be collected for DNA profiling and, in 2000, the Federal Government began requiring certain offenders convicted of Federal or military crimes to also provide a DNA sample for the criminal DNA database. Recognizing that the effectiveness of the DNA database relies on the volume of data contained in both the forensic index (crime scene samples) and the convicted offender index of CODIS, many States are changing their database statutes to include less violent criminals. Many States are enacting legislation to require all convicted felons to submit a DNA profile to the State database. The tendency for States to include all convicted felons in their databases dramatically increases the number of convicted offender DNA profiles against which forensic DNA evidence can be compared, thus making the database system a more powerful tool for law enforcement.

New legal approaches

DNA technology and DNA databases have encouraged the development of new approaches to old cases. One such approach is the filing of charges by "John Doe" warrant. These warrants are based on the unique DNA profile obtained from the analysis of unsolved crime scene evidence. Although John Doe warrants are traditionally filed based on the physical description or alias of an unnamed suspect, investigators and prosecutors are now filing charges using the suspect's DNA profile as the identifier. This innovative approach has allowed charges to be filed that toll and permit old cases to be prosecuted when the person matching the John Doe DNA profile is identified. John Doe DNA warrants are one way to permit cases to remain active, allowing them the chance to be solved through the DNA database in the future.

The Long and Short of DNA

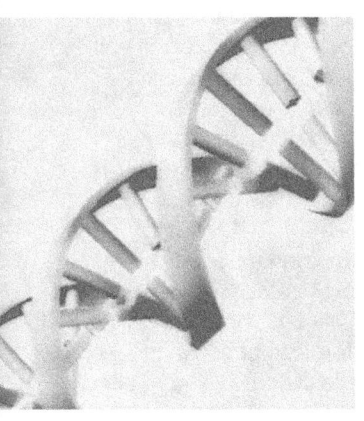

DNA is the fundamental building block for an individual's entire genetic makeup. It is a component of virtually every cell in the human body, and a person's DNA is the same in every cell. That is, the DNA in a person's blood is the same as the DNA in his skin cells, saliva, and other biological material.

DNA analysis is a powerful tool because each person's DNA is unique (with the exception of identical twins). Therefore, DNA evidence collected from a crime scene can implicate or eliminate a suspect, similar to the use of fingerprints. It also can analyze unidentified remains through comparisons with DNA from relatives. Additionally, when evidence from one crime scene is compared with evidence from another using CODIS, those crime scenes can be linked to the same perpetrator locally, statewide, and nationally.

DNA is also a powerful tool because when biological evidence from crime scenes is collected and stored properly, forensically valuable DNA can be found on evidence that may be decades old. Therefore, old cases that were previously thought unsolvable may contain valuable DNA evidence capable of identifying the perpetrator.

Similar to fingerprints

DNA is often compared with fingerprints in the way matches are determined. When using either DNA or fingerprints to identify a suspect, the evidence collected from the crime scene is compared with a "known" standard. If identifying features are the same, the DNA or fingerprint can be determined to be a match. However, if identifying features of the DNA profile or fingerprint are different from the known standard, it can be determined that it did not come from that known individual.

DNA technology advancements

Recent advancements in DNA technology have improved law enforcement's ability to use DNA to solve old cases. Original forensic applications of DNA analysis were developed using a technology called restriction fragment length polymorphism (RFLP). Although very old cases (more than 10 years) may not have had RFLP analysis done, this kind of DNA testing may have been attempted on more recent unsolved cases. However, because RFLP analysis required a relatively large quantity of DNA, testing may not have been successful. Similarly, biological evidence deemed insufficient in size for testing may not have been previously submitted for testing. Also, if a biological sample was degraded by environmental factors such as dirt or mold, RFLP analysis may have been unsuccessful at yielding a result. Newer technologies could now be successful in obtaining results.

Newer DNA analysis techniques enable laboratories to develop profiles from biological evidence invisible to the naked eye, such as skin cells left on ligatures or weapons. Unsolved cases should be evaluated by investigating both traditional and nontraditional sources of DNA. Valuable DNA evidence might be available that previously went undetected in the original investigation.

If biological evidence is available for testing or retesting in unsolved case investigations, it is important that law enforcement and the crime laboratory work together to review evidence. Logistical issues

If biological evidence is available for testing or retesting in unsolved case investigations, it is important that law enforcement and the crime laboratory work together to review evidence.

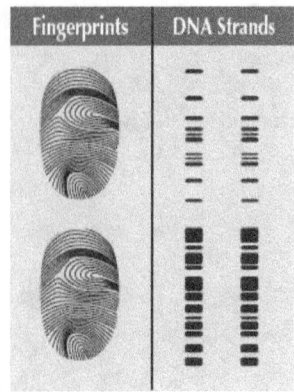

regarding access to and the cost of DNA analysis will be a factor, as well as issues that relate to the discriminating power of each technology and that might affect the outcome of the results. Laboratory personnel can also provide a valuable perspective on which evidence might yield valuable and probative DNA results. Finally, if previously tested biological evidence produced a DNA profile but excluded the original suspect, revisiting those "exclusion" cases in the context of comparing them with DNA databases might prove to be very valuable to solving old cases.

PCR analysis

PCR (polymerase chain reaction) enhances DNA analysis and has enabled laboratories to develop DNA profiles from extremely small samples of biological evidence. The PCR technique replicates exact copies of DNA contained in a biological evidence sample without affecting the original, much like a copy machine. RFLP analysis requires a biological sample about the size of a quarter, but PCR can be used to reproduce millions of copies of the DNA contained in a few skin cells. Since PCR analysis requires only a minute quantity of DNA, it can enable the laboratory to analyze highly degraded evidence for DNA. On the other hand, because the sensitive PCR technique replicates any and all of the DNA contained in an evidence sample, greater attention to contamination issues is necessary when identifying, collecting, and preserving DNA evidence. These factors may be particularly important in the evaluation of unsolved cases in which evidence might have been improperly collected or stored.

STR analysis

Short tandem repeat (STR) technology is a forensic analysis that evaluates specific regions (loci) that are found on nuclear DNA. The variable (polymorphic) nature of the STR regions that are analyzed for forensic testing intensifies the discrimination between one DNA profile and another. For example, the likelihood that any two individuals (except identical twins) will have the same 13-loci DNA profile can be as high as 1 in 1 billion or greater. The Federal Bureau of Investigation (FBI) has chosen 13 specific STR loci to serve as the standard for CODIS. The purpose of establishing a core set of STR loci is to ensure that all forensic laboratories can establish uniform DNA databases and, more importantly, share valuable forensic information. If the forensic or convicted offender CODIS index is to be used in the investigative stages of unsolved cases, DNA profiles must be generated by using STR technology and the specific 13 core STR loci selected by the FBI.

Mitochondrial DNA analysis

Mitochondrial DNA (mtDNA) analysis allows forensic laboratories to develop DNA profiles from evidence that may not be suitable for RFLP or STR analysis. While RFLP and PCR techniques analyze DNA extracted from the nucleus of a cell, mtDNA technology analyzes DNA found in a different part of the cell, the mitochondrion (see exhibit 1). Old remains and evidence lacking nucleated cells—such as hair shafts, bones, and teeth—that are unamenable to STR and RFLP testing may yield results if mtDNA analysis is performed. For this reason, mtDNA testing can be very valuable to the investigation of an unsolved case. For example, a cold case log may show that biological evidence in the form of blood, semen, and hair was collected in a particular case, but that all were improperly stored for a long period of time. Although PCR analysis sometimes enables the crime laboratory to generate a DNA profile from very degraded evidence, it is possible that the blood and semen would be so highly degraded that nuclear DNA analysis would not yield a DNA profile. However, the hair

If the convicted offender or forensic index of CODIS is to be used in the investigative stages of an unsolved case, DNA profiles must be generated using STR analysis.

shaft could be subjected to mtDNA analysis and thus be the key to solving the case. Finally, it is important to note that all maternal relatives (for example, a person's mother or maternal grandmother) have identical mtDNA. This enables unidentified remains to be analyzed and compared to the mtDNA profile of any maternal relative for the purpose of aiding missing persons or unidentified remains investigations. Although mtDNA analysis can be very valuable to the investigation of criminal cases, laboratory personnel should always be involved in the process.

Exhibit 1. Cell diagram

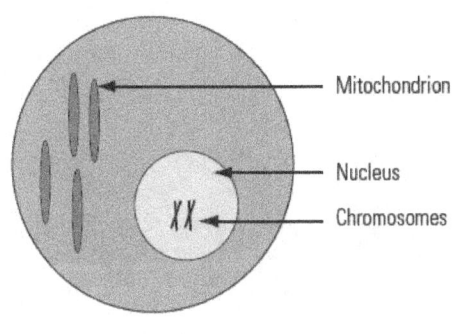

Y-chromosome analysis

Several genetic markers have been identified on the Y chromosome that can be used in forensic applications. Y-chromosome markers target only the male fraction of a biological sample. Therefore, this technique can be very valuable if the laboratory detects complex mixtures (multiple male contributors) within a biological evidence sample. Because the Y chromosome is transmitted directly from a father to all of his sons, it can also be used to trace family relationships among males. Advancements in Y-chromosome testing may eventually eliminate the need for laboratories to extract and separate semen and vaginal cells (for example, from a vaginal swab of a rape kit) prior to analysis.

Cooperative efforts with the crime laboratory are essential to deciding which analysis methods will be most valuable in a particular case. It is important to note, however, that while RFLP and mtDNA testing may be valuable to the investigation of an old case, current DNA databases are being populated with DNA profiles that are generated using STR analysis. RFLP and mtDNA profiles are not compatible with the convicted offender or forensic indexes of CODIS.[2]

How Can DNA Databases Aid Investigations?

The development and expansion of databases that contain DNA profiles at the local, State, and national levels have greatly enhanced law enforcement's ability to solve cold cases with DNA. Convicted offender databases store hundreds of thousands of potential suspect DNA profiles, against which DNA profiles developed from crime scene evidence can be compared.

Success Story

A "forensic hit" occurred in the National DNA Index System (NDIS) that linked a dead Florida man's DNA profile to eight serial unsolved rapes in Washington, D.C. and three offenses in Florida.

In 1999, Leon Dundas was killed in a drug deal. Investigators remembered Dundas refusing to give a blood sample in connection with a rape investigation in 1998. They were able to obtain Dundas' blood sample through the medical examiner's office and forwarded it to the DNA lab at the Florida Department of Law Enforcement. Dundas' DNA profile was compared with the national forensic index and a match was made between Dundas and DNA evidence from a rape victim in Washington, D.C.

The FBI then entered DNA evidence from additional unsolved rapes committed in Washington. Dundas' DNA matched seven additional rapes in Washington and three more in Jacksonville, Florida. Police in Washington said that without DNA, they would have never identified Dundas, who had no prior recorded history of violent crime.

Given the recidivistic nature of many crimes, such as sexual assault and burglary, a likelihood exists that the individual who committed the crime being investigated was convicted of a similar crime and already has his or her DNA profile in a DNA database that can be searched by CODIS. Moreover, CODIS also permits the cross-comparison of DNA profiles developed from biological evidence found at crime scenes. Even if a perpetrator is not identified through the database, crimes may be linked to each other, thereby aiding an investigation, which may eventually lead to the identification of a suspect.

What is CODIS?

CODIS is a computer software program that operates local, State, and national databases of DNA profiles from convicted offenders, unsolved crime scene evidence, and missing persons. Every State in the Nation has a statutory provision for the establishment of a DNA database that allows for the collection of DNA profiles from offenders convicted of particular crimes. CODIS software enables State, local, and national law enforcement crime laboratories to compare DNA profiles electronically, thereby linking serial crimes to each other and identifying suspects by matching DNA profiles from crime scenes with profiles from convicted offenders. The success of CODIS is demonstrated by the thousands of matches that have linked serial cases to each other and cases that have been solved by matching crime scene evidence to known convicted offenders.

The missing persons index consists of the unidentified persons index and the reference index. The unidentified persons index contains DNA profiles from recovered remains, such as bone, teeth, or hair. The reference index contains DNA profiles from related individuals of missing persons so that they can be periodically compared to the unidentified persons index. All samples for this index are typed using mtDNA and STR DNA analysis (if possible) to maximize the power of advancing technology.

SPECIAL REPORT / JULY 02

The offender index contains DNA profiles of individuals convicted of certain crimes. The forensic index contains DNA profiles obtained from crime scene evidence.

How does CODIS work?

CODIS uses two indexes to generate investigative leads in crimes for which biological evidence is recovered from a crime scene. The convicted offender index contains DNA profiles of individuals convicted of certain crimes ranging from certain misdemeanors to sexual assault and murder. Each State has different "qualifying offenses" for which persons convicted of them must submit a biological sample for inclusion in the DNA database. The forensic index contains DNA profiles obtained from crime scene evidence, such as semen, saliva, or blood. CODIS uses computer software to automatically search across these indexes for a potential match.

A match made between profiles in the forensic index can link crime scenes to each other, possibly identifying serial offenders. Based on these "forensic hits," police in multiple jurisdictions or States can coordinate their respective investigations and share leads they have developed independent of each other. Matches made between the forensic and convicted offender indexes can provide investigators with the identity of a suspect(s). It is important to note that if an "offender hit" is obtained, that information typically is used as probable cause to obtain a new DNA sample from that suspect so the match can be confirmed by the crime laboratory before an arrest is made.

LDIS, SDIS, and NDIS

CODIS is implemented as a distributed database with three hierarchical levels (or tiers)—local, State, and national. All three levels contain forensic and convicted offender indexes and a population file (used to generate statistics). The hierarchical design provides State and local laboratories with the flexibility to configure CODIS to meet their specific legislative and technical needs.

A description of the three CODIS tiers follows (see exhibit 2).

- Local. Typically, the Local DNA Index System (LDIS) installed at crime laboratories is operated by police departments or sheriffs' offices. DNA profiles originated at the local level can be transmitted to the State and national levels.

- State. Each State has a designated laboratory that operates the State DNA Index System (SDIS). SDIS allows local laboratories within that State to compare DNA profiles. SDIS also is the communication path between the local and national tiers. SDIS is typically operated by the agency responsible for implementing and monitoring compliance with the State's convicted offender statute.

- National. The National DNA Index System (NDIS) is the highest level of the CODIS hierarchy and enables qualified State laboratories that are actively participating in CODIS to compare DNA profiles. NDIS is maintained by the FBI under the authority of the DNA Identification Act of 1994.

Limitations of using the DNA database

The more data contained in the forensic and offender indexes of CODIS, the more powerful a tool it becomes for law enforcement, especially in its application to unsolved case investigation. However, because many jurisdictions are in the process of developing and populating their DNA databases, convicted offender and forensic casework backlogs have been created over time and continue to grow for several reasons. First, as States recognize the crime-solving potential of DNA databases, they continue to expand the scope of their convicted offender legislation, which increases the number of

Exhibit 2. CODIS tiers

samples to be collected and analyzed by the DNA laboratory. As a result, more than 1 million uncollected convicted offender DNA profiles are "owed" to the system.

An equally important but more difficult problem to quantify is that of unprocessed casework that contains biological evidence. This casework backlog may include nonsuspect or unsolved cases that could be analyzed and solved as a result of advancements in DNA technology.

Convicted offender backlogs

Although all 50 States have passed DNA database legislation, many States have backlogs of convicted offender samples that have been collected but have not yet been analyzed. Although Federal funding has played an important role in reducing existing backlogs, the crimefighting potential of DNA has prompted many States to revise their statutes to require nonviolent convicted offenders to provide a DNA sample for analysis and upload into CODIS. The trend toward expanding convicted offender DNA statutes to include nonviolent offenders has significantly increased the number of DNA samples requiring collection and analysis. Although the success of using the DNA database as a crime-solving and crime-prevention tool can easily be demonstrated once convicted offender backlogs are reduced, it should be recognized that new backlogs are instantly created by the passage of expanded DNA legislation laws. Convicted offender backlogs are an ongoing logistical issue that can compound the complexity of investigating cold cases by using the DNA database.

Forensic casework backlogs

Addressing issues that affect the efficient and effective use of DNA databases in the United States is complicated further by the existence of casework backlogs. This refers to biological evidence in perhaps tens of thousands of criminal cases, including violent and nonviolent crimes, that has not been tested or retested for DNA.

Unprocessed rape kits are a clear example of this kind of backlog. Despite the established fact that rape typically yields biological evidence, as of October 1999, at least 180,000 rape kits remained on shelves across the country, unprocessed, because no suspects have been identified. The DNA evidence from these and other criminal cases often is not analyzed and entered into the DNA database because forensic laboratories have to prioritize their work and cases scheduled for trial take precedence over cases in which no suspect is known. In most jurisdictions, nonsuspect criminal cases that contain biological evidence are not being analyzed and entered into the DNA database. In many jurisdictions, DNA from crime scenes is still primarily used to prosecute offenders, not to investigate crimes. The convicted offender backlog and limited resources for casework going to trial preclude State forensic laboratories from analyzing all biological evidence for DNA, which in turn prevents law enforcement from being able to realize the full crime-solving potential of CODIS.

The backlog of forensic cases has practical consequences for most law enforcement agencies in the United States. Laboratory capacity limitations result in the ability to process crime scene samples from only the most serious of offenses. More and more, however, agencies such as those in the United Kingdom are discovering the value of DNA technology in solving property crimes. Blood left on a broken apartment window or saliva found on a discarded beer bottle can be used to identify burglars, and the skin cells rubbed off onto the steering wheel of a stolen vehicle can solve car thefts. However, as long as forensic laboratories remain able to process only the most serious cases, the full potential of DNA technology to solve crime will remain untapped.

Practical Considerations

A broad range of considerations must be made long before any DNA testing is actually attempted in older, unsolved cases. These include—

- Legal considerations, such as the application or expiration of statutes of limitation.

- Technological considerations, such as the nature and condition of the evidence as originally collected, stored, and in some instances, subjected to other forensic tests.

- Practical considerations, such as the availability of witnesses in the event DNA testing would identify a suspect and lead to an arrest and a trial.

- Resource issues, such as the time and money available for investigation and forensic analysis.

The nature and scope of these issues require that any approach to reexamining old cases for potential DNA evidence be collaborative, whether by an individual investigator or by a specialized unit developed specifically for cold case review. Local prosecutors can provide valuable insight into legal issues that might prevent or help a future prosecution. Victim/witness units or advocates can provide valuable assistance with locating, educating, and encouraging witnesses. Consultation with representatives from the crime laboratory is critical to ensuring that potential DNA evidence can be successfully analyzed.

Local prosecutors can provide valuable insight into legal issues. Victim/witness units or advocates can help locate, educate, and encourage witnesses. Consultation with representatives from the crime laboratory is critical.

Evidence considerations

When collecting unsolved case evidence from storage facilities, the case investigator should be ready to handle all types of packaging disasters. Evidence may be stored in heavy-duty plastic bags, stapled shut as the past form of "sealing." Multiple items may be sealed in one plastic bag, or even unpackaged in large, open, cardboard boxes. Unprotected microscope slides from medical facilities might also be found as a result of investigating old cases. No attempt should be made on the part of the investigator to separate and repackage evidence. The condition and position that the evidence has been stored in could provide valuable clues to the forensic scientist for testability of evidence. Only when evidence is found unpackaged should the investigator properly package and label the item(s) to minimize the possibility for contamination from that point forward. It is important that any evidence items are handled minimally and only by individuals wearing disposable gloves. As always, it is also very important that all actions taken as a result of opening, evaluating, packaging, or repackaging evidence are documented thoroughly in the case folder.

Degraded evidence

Prior to the frequent use of DNA technology, biological evidence may have been collected and stored in ways that were not necessarily the best methods for preserving samples for future DNA testing. For example, evidence containing biological fluids that were originally collected for ABO Blood Typing analysis or other serology methods may have been packaged or stored in ways that can limit DNA testing. Some methods of collection and storage may promote the growth of bacteria and mold on the evidence. Bacteria can seriously damage or degrade DNA contained in biological material and inhibit the ability to develop a DNA profile; however, evidence can still sometimes yield DNA results. For example, PCR technology can allow the laboratory to develop profiles

from some moldy biological samples, whereas other evidence may fail to yield a usable DNA profile, even when no mold is visible. Therefore, close consultation with the laboratory is important to determine the type of DNA testing most likely to yield results on the available evidence.

Contamination issues

Because of the particularly sensitive nature of DNA technology, the potential contamination of evidence should be carefully considered. Technologies used to analyze evidence prior to the forensic application of DNA were not always sensitive to contaminants. Evidence in older cases may have been collected in ways that lacked appropriate contamination or cross-contamination safeguards, which can make the DNA results less useful or even misleading. In these cases, clarifying results by identifying the contributor of an additional profile can determine whether the DNA results may now be used. When a mixture is detected, a careful reconstruction of the evidence collection, storage, and analysis process must be undertaken. It may be determined that DNA profiles will be required from on-scene officers, evidence technicians, or laboratory scientists who had access to the evidence for comparison with evidence results. In these instances, proper chain-of-custody reconstruction is critical.

EVIDENCE HANDLING RECOMMENDATIONS

- Wear gloves. Change them between handling each item of evidence.
- Use disposable instruments or clean instruments thoroughly before and after handling each evidence sample.
- Avoid touching the area where you believe DNA may exist.
- Avoid touching your face, nose, and mouth when examining and repackaging evidence.
- Put dry evidence into new paper bags or envelopes; do not use plastic bags.
- Do not use staples.
- If repackaging of evidence is necessary, consult with laboratory personnel.

It is also important to avoid contamination when handling biological evidence during the course of the current review. If evidence that may contain biological material is already sealed, do not reopen it before sending it to the laboratory. (See Evidence Handling Recommendations.)

Legal considerations

Numerous legal issues might arise when examining older cases for potential DNA evidence. These issues are most likely jurisdictionally specific and may differ from State to State. Although most jurisdictions maintain no statute of limitation for filing charges in a homicide case, States can vary widely in the time allowed for filing charges in other cases, such as rape and other sexual assault crimes. Furthermore, in recognition of DNA technology's ability to solve old cases, many States are extending or even eliminating statutes of limitation for certain crimes.

Chain of custody

When a case remains unsolved for a long period of time, evidence is usually handled by an increased number of individuals. Many unsolved cases to be reviewed for DNA evidence may have been previously reinvestigated or handled by several different investigators as a result of new leads or periodic, systematic reviews. Furthermore, as cases age, the likelihood increases that evidence may be moved to new or remote storage locations as evidence from newer cases fills police department shelves.

Many cases may also have had evidence submitted to the laboratory for various forms of forensic testing. Evidence in older cases may have been submitted for standard serological testing, but can now be tested for DNA with much greater success. Hair previously submitted for standard microscopic hair analysis may now be submitted for mtDNA testing. As with all criminal investigations, chain-of-custody issues are critical to maintaining the integrity of the evidence. In all cases, the ultimate ability to use DNA evidence will depend on the ability to prove that the chain of custody was maintained.

Statutes of limitation

One of the first issues to address when reviewing an unsolved case is whether the statutes of limitation on a case have run out. Several considerations arise when addressing a statute of limitation issue. Good communication between law enforcement and local prosecutors is critical when examining these legal questions.

Changes in statutes. Advances in DNA technology and the creation of DNA databases are leading many criminal justice professionals to rethink time limits placed on the filing of criminal charges. Because biological evidence can yield reliable DNA analysis results years after the commission of a crime, many State legislatures have begun to extend, and in some cases eliminate, the statutes of limitation for some crimes and in certain circumstances. Many States have extended the length of time for which a complaint can be filed, other States have eliminated statutes of limitation for certain crimes, and some legislation is retroactive.

Exceptions to statutes. Exceptions often exist under existing and new statutes. Under such exceptions, time can be added to the statute of limitation, giving police the legal authority to arrest even if it appears as though the statute has run out. For example, many jurisdictions have exceptions for a suspect's flight from jurisdiction. In a case for which there is a 5-year statute of limitation, if the government can prove that the suspect has been absent from the jurisdiction for 2 years, the State can still file against the suspect for up to 7 years after the commission of the crime. Exceptions also exist for cases in which child victims are assaulted by a family member, which can be valuable in the context of a current investigation.

Victim and witness considerations

Another important consideration to be made early in the process is the willingness of victims and witnesses to proceed. Although many victims may continuously monitor the progress of their investigations, some choose to detach from the process over time. Reinvestigating a case may cause renewed psychological trauma to the victim and victim's family. It should not be assumed that victims and witnesses, even if they were eager to pursue the case when it occurred, are still interested in pursuing the case. A phone call from an investigator years later may not be a welcome event. Whenever possible, enlist the aid of victim service providers. If a new officer is handling the investigation, enlisting the assistance of the original investigator to make the first contact with the victim may also be helpful.

The older a case is, the more difficult it may be to locate witnesses. However, early identification of victim and witness availability may ultimately save significant resources. Consultation with prosecutors is mandatory when considering whether a witness would be necessary at trial.

It should not be assumed that victims and witnesses are still interested in pursuing the case. Whenever possible, enlist the aid of victim service providers.

STATUTE OF LIMITATION RECOMMENDATIONS

- Know the original statute of limitation.
- Determine whether the law has changed regarding time limits for filing. If so, is the law retroactive?
- Determine whether there are exceptions to the statute.
- Consult with the prosecutor.

Identifying, Analyzing, and Prioritizing Cases

Good communication between police, laboratories, and prosecutors can help identify and convict serious offenders and save valuable time and resources.

Whether the process of reviewing unsolved cases is initiated by a single officer or by a specialized unit, it must ultimately be a team effort. At all stages of the process, investigators should avail themselves of the scientific advice of the laboratory and the legal expertise of the local prosecutor's office. Close consultation with the laboratory can ensure that evidence integrity is maintained and that limited laboratory resources are allocated effectively. Similarly, prosecutors can help identify issues that might occur at trial if a suspect is identified and arrested upon successful DNA testing. Good communication between police, laboratories, and prosecutors can help identify and convict serious offenders and save valuable time and resources.

Identify potential cases for review

An initial step in the DNA review of unsolved cases is to identify cases that might be amenable to DNA testing. While the cases considered for this kind of review will vary from jurisdiction to jurisdiction, it is important to define minimum requirements that will likely benefit from this approach. Issues such as statutes of limitation and solvability factors should be thoroughly examined in cooperation with a prosecutor and the forensic laboratory to establish guidelines for case selection. It also will be important to identify the ultimate goals of the program so that the selection criteria can be tailored to meet those specific goals.

Cases that could benefit from a review for potential DNA evidence can be identified from numerous sources. In some instances a single police officer or investigator may remember an unsolved case from years ago. In some departments a formalized cold-case unit may systematically review cases for the potential of DNA testing. Other cases may be identified by coordinated, interdepartmental efforts, victims or witnesses who have heard about the potential of DNA evidence, and laboratories taking inventory of their storage facilities. If a department is pursuing a systematic review of cases, either by one or two officers or by a formal unit, there are many sources that can be consulted for valuable investigative information, such as—

- Autopsy, laboratory, prosecutor, and local agency logbooks.
- Retired investigators.
- Computer databases.

Identify statute of limitation issues

Statute of limitation issues might affect the ultimate ability to prosecute a case. Cases should be preliminarily reviewed by investigators in conjunction with the prosecutor's office to identify which prosecutions would be barred by the statutes of limitation. If the goal of the unsolved case review program is to obtain convictions and statutes of limitation have expired on a particular case, a department may wish to save its resources for cases likely to yield convictions. However, if the goal of the program is to solve and close unsolved cases regardless of whether a conviction could be obtained, a jurisdiction may decide to review all cases that qualify under its guidelines. This is an important consideration in the context of investigating serial offenders whose criminal acts might span the course of years or decades.

Define categories of cases—solvability factors

Because the number of cases that qualify for reinvestigation might be very large, it may be beneficial for a jurisdiction to define cases according to several solvability factors. Solvability factors include facts and circumstances of a case that influence the likelihood that it might be solved through advancements in DNA technology. For example, a high probability exists that analysis of nonsuspect rape kits will yield valuable DNA results. Profiles generated as a result of DNA analysis can now be entered into CODIS, which can solve a case by matching to a convicted offender, or aid investigations by linking serial rapes to each other. Additionally, if an unsolved murder case contains biological evidence foreign to the victim that did not produce viable results from ABO blood typing or RFLP DNA analysis, evidence could be reanalyzed with the more discriminating and powerful STR technology. It is also important to recognize and sort out cases that might not be as likely to be solved with DNA technology. An example might be an unsolved drive-by homicide because the perpetrator most likely would not have left biological evidence at this kind of crime scene.

Case review—establish priorities

Once solvability factors and statute of limitation issues are addressed, it is important to continue the process by identifying the cases to be reviewed first. To preserve investigative resources when considering a larger number of unsolved cases for review, jurisdictions may prioritize according to the likelihood that cases will be solved or the likelihood that investigations will be aided. In establishing this priority, the following criteria can be considered:

- How many qualifying cases are there?
- Where are the case files located?
- Are case summaries available?
- How many cases will be assigned to an investigator?

To establish an investigative hierarchy, qualifying cases should be reviewed by experienced, proficient investigators. A checklist can be used throughout the review process so that managers can decide which cases will be worked first. A checklist can also provide review process consistency throughout the agency. (See Sample Checklist at the end of this report.) The following categories may serve as a model for a hierarchy in prioritizing cases:

- There is a known suspect and physical evidence appears to have been preserved in a manner consistent with successful DNA testing and use of CODIS.

- There is no known suspect but physical evidence has been preserved in a manner consistent with successful DNA testing and use of CODIS.

- There is no known suspect and evidence was collected and preserved in a manner that may make it difficult to obtain a DNA profile.

Locating case files, obtaining evidence logs, and other documentation

Locating the case file and original evidence for the investigation may be a challenging endeavor. Changes in personnel, procedure, and facilities and the passage of time may complicate the process. When searching for a case file or evidence, an investigator may need to look in numerous

USING DNA TO SOLVE COLD CASES

places. Potential locations include, but are not limited to, the following:

- Police department property rooms (case files, evidence logs, whole evidence).
- Property warehouses (case files, evidence logs, whole evidence).
- Public crime laboratories (previously tested/submitted evidence, lab reports).
- Private laboratories (previously tested evidence, lab reports).
- Hospital/medical facilities (rape kits, medical reports, slides).
- Coroner/medical examiners' offices (autopsy reports).
- Courthouse property rooms.
- Prosecutors' offices (previous trial or suspect investigation).
- Retired investigators' files (case notes and details not contained in file).
- Other investigating agency offices (investigative leads—serial offender).

Forensic testing reports and previously tested evidence

Because advancements in DNA technology enable laboratories to successfully analyze old evidence that might have been improperly stored or subjected to previous forensic analysis, it will be very valuable to locate any and all forensic reports that were produced as a result of previous analysis and/or testing. ABO blood typing, microscopic hair analysis, RFLP DNA analysis, or fingerprint analysis (among others) might have been performed in the course of the original investigation. The original case file should indicate whether and which types of forensic analysis were attempted. These reports also serve to memorialize proper chain of custody. Cooperation with the crime laboratory is crucial to locate and interpret existing forensic reports and to determine whether evidence would be amenable to reanalysis with new DNA techniques.

Many combinations of options are available to investigators and laboratory personnel if biological evidence was available and previously tested. Exhibit 3 may serve to help investigators as they work with the laboratory to discuss options throughout the course of the investigation.

Locate biological evidence

When reviewing the case file for potential DNA evidence, it is important to know what kinds of evidence may yield a DNA profile. Given the power and sensitivity of newer DNA testing techniques, DNA can be collected from virtually anywhere. Only a few cells can be sufficient to obtain useful DNA information to help solve a case. Exhibit 4 identifies some common items of evidence that may have been collected previously but not analyzed for the presence of DNA evidence. Remember, if a stain is not visible it does not mean that there are not enough cells for DNA typing. Further, DNA does more than just identify the source of the sample; it can place a known individual at a crime scene, in a home, or in a room where the suspect claimed not to have been. It can refute a claim of self-defense and put a weapon in the suspect's

DNA CAN DO MORE...

...than identify a suspect. It can also—
- Place a known individual at a crime scene.
- Refute a claim of self-defense.
- Put a weapon in a suspect's hand.
- Change a suspect's story from an alibi to one of consent.

hand. It can also provide irrefutable evidence that can change a suspect's story from an alibi to one of consent.

Evaluate for probative DNA evidence

On completion of reviewing the case file, reports, and evidence in consultation with the laboratory, it will be necessary to identify which evidentiary items will be amenable to DNA analysis. Consultation with the laboratory will be essential to determine the likelihood of obtaining results from DNA analysis, and consultation with a prosecutor is very important to determine which evidence will be probative to the case. Building the new investigation on cooperative efforts between the laboratory and prosecutor can save valuable resources, develop leads, and identify previously overlooked evidence that may yield a DNA profile.

Continue investigative protocol

If DNA analysis is to be conducted, it may be important to obtain reference samples from prior suspects, and it might be necessary to be creative when obtaining these samples. While a biological sample in the form of blood or saliva can be obtained voluntarily through a consent form, a standard reference sample might already exist if previous forensic analysis,

Exhibit 3. Investigative options

Test conducted	Original results	Original interpretation	Options for investigators
RFLP/PCR	Obtained profiles.	No suspects identified.	Is the original extract remaining? 1. If so, retest using STR technique and submit to CODIS. 2. If not, reextract the original sample using STR technique and submit to CODIS.
RFLP	Inconclusive or no results obtained.	Sample size may have been insufficient or not concentrated enough.	Is the original extract remaining? 1. If so, retest using STR technique and submit to CODIS. 2. If not, reextract the original sample using STR technique and submit to CODIS.
PCR	Inconclusive or result intensity below "S" and "C" dots.	Sample size may have been insufficient.	Is the original extract remaining? 1. If so, retest using STR technique and submit to CODIS. 2. If not, reextract the original sample using STR technique and submit to CODIS.
Conventional serology (ABO, secretor status, enzymes such as EsD, PGM, GLO I, EAP, ADA, AK).	Obtained a type in these systems.	Poor statistics and no searching capability.	If original evidence still exists, extract the sample using STR technique and submit to CODIS.
None		1. Limited sample size. 2. No suspects, did not process further. 3. No request at the time of analysis.	If original evidence still exists, extract the sample using STR technique and submit to CODIS.

USING DNA TO SOLVE COLD CASES

such as serological testing, was performed during the course of the original investigation.

Additionally, elimination samples from anyone who had lawful access to the crime scene, such as family members, may be required if the laboratory determines that there is more than one DNA profile present in the evidence sample. Early identification of the location and status of persons who might be requested to submit an elimination sample could save valuable time and resources if the laboratory needs such information. Consultation with the laboratory is essential to properly coordinating this process.

Follow agency procedures for submitting the DNA profile to CODIS

On successful laboratory analysis resulting in a DNA profile developed from crime scene evidence, existing and/or new suspect DNA profiles should be compared with the evidence profile. If the laboratory determines a match between a suspect and the evidence, the prosecutor's office should be consulted on how to proceed. However, if a match is not found, agency procedures should be followed, in accordance with the crime laboratory, to submit the crime scene evidence DNA profile into CODIS.

Exhibit 4. Common items of evidence

Evidence	Possible location of DNA on the evidence	Source of DNA
Baseball bat	Handle	Skin cells, sweat, blood, tissue
Hat, bandanna, or mask	Inside surfaces	Sweat, hair, skin cells, dandruff, saliva
Eyeglasses	Nose or ear piece, lens	Sweat, skin cells
Facial tissue, cotton swab	Surface	Mucus, blood, sweat, semen, ear wax
Dirty laundry	Surface	Blood, sweat, semen, saliva
Toothpick	Surface	Saliva
Used cigarette	Cigarette butt (filter area)	Saliva
Used stamp/envelope seal	Moistened area	Saliva
Tape or ligature	Inside or outside surface	Skin cells, sweat, saliva
Bottle, can, or glass	Mouthpiece, rim, outer surface	Saliva, sweat, skin cells
Used condom	Inside/outside surface	Semen, vaginal or rectal cells
Bed linens	Surface	Sweat, hair, semen, saliva, blood
"Through and through" bullet	Outside surface	Blood, tissue
Bite mark	Surface of skin	Saliva
Fingernail/partial fingernail	Scrapings	Blood, sweat, tissue, skin cells

Note: When reviewing evidence, it is important to maintain chain of custody, consult with laboratory personnel, and take all appropriate precautions against contamination, including wearing gloves and changing them between handling of different pieces of evidence.

Because CODIS contains hundreds of thousands of convicted offender DNA profiles, it is possible that the person who committed the unsolved crime being investigated was convicted of a qualifying offense that required submission of a DNA profile to the database. If that person has not previously been convicted of a qualifying offense, especially in light of expanding database law, it is possible that they will be convicted in the future. Further, because the forensic index of CODIS contains thousands of crime scene evidence profiles, the investigation could be aided if a match is made to another forensic DNA profile already in the database. Finally, an investigator should not assume that a new DNA profile generated from unsolved case evidence and submitted to the laboratory for entry into CODIS will be compared with every possible convicted offender or crime scene index profile. The investigator may need to proactively request that his CODIS administrator search the new profile against the local, State, and national DNA databases.

Prepare a John Doe warrant

CODIS is a powerful crime-solving and crime-prevention tool, but many cases will not be solved as a result of entering a DNA profile into the forensic index of the database. Additionally, many cases will have statute of limitation issues that might prevent the prosecution of the case if a match is not determined in a timely manner. Therefore, if no offender match occurs in cases in which statutes of limitation are an issue, consideration may be given, in consultation with the prosecutor, to preparing a John Doe warrant. These types of warrants can identify the perpetrator according to his or her DNA profile. The 13-loci profile generated by the crime laboratory should be clearly printed on the face of the warrant. The John Doe warrant is not novel; however, the unconventional method of describing an individual by his or her DNA profile may allow for prosecution of a case if a DNA match is determined in the course of future investigations or as a result of the CODIS system being populated with more convicted offender and forensic DNA profiles.

Notes

1. CODIS uses two indexes—the forensic index and the offender index—to generate investigative leads in crimes where biological evidence is recovered from crime scenes. The forensic index contains DNA profiles of biological crime scene evidence and the offender index contains DNA profiles of individuals convicted of a qualifying offense.

2. CODIS has a missing persons index that exclusively contains mtDNA profiles; the convicted offender and forensic indexes of CODIS exclusively contain STR DNA profiles.

USING DNA TO SOLVE COLD CASES

SAMPLE CHECKLIST

- Identify potential cases.
  - Identify any statute of limitation issues (consult with prosecutors).
  - Define case categories according to solvability factors.

- Prioritize cases (consider solvability factors).

- Locate and review the case file; obtain evidence logs and other documentation such as laboratory and autopsy reports.

- Locate previous forensic testing reports and location of previously tested evidence. For example—
  - Blood previously ABO typed.
  - Hair analyzed microscopically.
  - Fingerprint evidence.

- Locate crime scene evidence containing biological material.

- Evaluate the case and evidence for potential probative DNA. Be sure to—
  - Consider all evidentiary possibilities.
  - Take appropriate precautions against contamination.

- In consultation with the laboratory and prosecutors, submit appropriate (probative) evidence to the laboratory for testing.

- Continue investigative protocol. If needed, obtain reference samples from suspects—
  - Voluntarily using a consent form.
  - By using a previously obtained sample (e.g., if a reference sample was used for standard serological testing).

- Identify witness issues—
  - Legal availability.
  - Willingness to proceed.
  - Location.

- If a profile does not match suspect profiles, follow agency procedures for submitting the evidence profile to CODIS.

- If no offender match occurs in cases in which statutes of limitation are an issue, prepare a John Doe warrant.

About the National Institute of Justice

NIJ is the research, development, and evaluation agency of the U.S. Department of Justice and is solely dedicated to researching crime control and justice issues. NIJ provides objective, independent, nonpartisan, evidence-based knowledge and tools to meet the challenges of crime and justice, particularly at the State and local levels. NIJ's principal authorities are derived from the Omnibus Crime Control and Safe Streets Act of 1968, as amended (42 U.S.C. §§ 3721–3722).

NIJ's Mission

In partnership with others, NIJ's mission is to prevent and reduce crime, improve law enforcement and the administration of justice, and promote public safety. By applying the disciplines of the social and physical sciences, NIJ—

- Researches the nature and impact of crime and delinquency.
- Develops applied technologies, standards, and tools for criminal justice practitioners.
- Evaluates existing programs and responses to crime.
- Tests innovative concepts and program models in the field.
- Assists policymakers, program partners, and justice agencies.
- Disseminates knowledge to many audiences.

NIJ's Strategic Direction and Program Areas

NIJ is committed to five challenges as part of its strategic plan: 1) *rethinking* justice and the processes that create just communities; 2) *understanding* the nexus between social conditions and crime; 3) *breaking* the cycle of crime by testing research-based interventions; 4) *creating* the tools and technologies that meet the needs of practitioners; and 5) *expanding* horizons through interdisciplinary and international perspectives. In addressing these strategic challenges, the Institute is involved in the following program areas: crime control and prevention, drugs and crime, justice systems and offender behavior, violence and victimization, communications and information technologies, critical incident response, investigative and forensic sciences (including DNA), less-than-lethal technologies, officer protection, education and training technologies, testing and standards, technology assistance to law enforcement and corrections agencies, field testing of promising programs, and international crime control. NIJ communicates its findings through conferences and print and electronic media.

NIJ's Structure

The NIJ Director is appointed by the President and confirmed by the Senate. The NIJ Director establishes the Institute's objectives, guided by the priorities of the Office of Justice Programs, the U.S. Department of Justice, and the needs of the field. NIJ actively solicits the views of criminal justice and other professionals and researchers to inform its search for the knowledge and tools to guide policy and practice.

NIJ has three operating units. The Office of Research and Evaluation manages social science research and evaluation and crime mapping research. The Office of Science and Technology manages technology research and development, standards development, and technology assistance to State and local law enforcement and corrections agencies. The Office of Development and Communications manages field tests of model programs, international research, and knowledge dissemination programs. NIJ is a component of the Office of Justice Programs, which also includes the Bureau of Justice Assistance, the Bureau of Justice Statistics, the Office of Juvenile Justice and Delinquency Prevention, and the Office for Victims of Crime.

To find out more about the National Institute of Justice, please contact:

National Criminal Justice
 Reference Service
P.O. Box 6000
Rockville, MD 20849–6000
800–851–3420
e-mail: *askncjrs@ncjrs.org*

www.ingramcontent.com/pod-product-compliance
Lightning Source LLC
Chambersburg PA
CBHW081813170526
45167CB00008B/3421